"An American Hero"
Davy Crockett

大衞克羅傳

U0061847

商務印書館

出版說明

　　本館一向倡導優質閱讀，近年來連續推出了以“Q”為標識的“Quality English Learning 優質英語學習”系列，其中《讀名著學英語》叢書，更是香港書展入選好書，讀者反響令人鼓舞。推動社會閱讀風氣，推動英語經典閱讀，藉閱讀拓廣世界視野，提高英語水平，已經成為一種潮流。

　　然良好閱讀習慣的養成非一日之功，大多數初、中級程度的讀者，常視直接閱讀厚重的原著為畏途。如何給年輕的讀者提供切實的指引和幫助，如何既提供優質的學習素材，又提供名師的教學方法，是當下社會關注的重要問題。針對這種情況，本館特別延請香港名校名師，根據多年豐富的教學經驗，精選海外適合初、中級英語程度讀者的優質經典讀物，有系統地出版了這套叢書，名為《Black Cat 優質英語階梯閱讀》。

　　《Black Cat 優質英語階梯閱讀》體現了香港名校名師堅持經典學習的教學理念，以及多年行之有效的學習方法。既有經過改寫和縮寫的經典名著，又有富創意的現代作品；既有精心設計的聽、說、讀、寫綜合練習，又有豐富的歷史文化知識；既有彩色插圖、繪圖和照片，又有英美專業演員朗讀作品的 CD。適合口味不同的讀者享受閱讀之樂，欣賞經典之美。

　　《Black Cat 優質英語階梯閱讀》由淺入深，逐階提升，好像參與一個尋寶遊戲，入門並不難，但要真正尋得寶藏，需要投入，更需要堅持。只有置身其中的人，才能體味純正英語的魅力，領略得到真寶的快樂。當英語閱讀成為自己生活的一部分，英語水平的提高自然水到渠成。

商務印書館（香港）有限公司

編輯部

使用説明 ─────────────

❶ 應該怎樣選書？

按閱讀興趣選書

《Black Cat 優質英語階梯閱讀》精選世界經典作品，也包括富於創意的現代作品；既有膾炙人口的小説、戲劇，又有非小説類的文化知識讀物，品種豐富，內容多樣，適合口味不同的讀者挑選自己感興趣的書，享受閱讀的樂趣。

按英語程度選書

《Black Cat 優質英語階梯閱讀》現設 Level 1 至 Level 6，由淺入深，涵蓋初、中級英語程度。讀物分級採用了國際上通用的劃分標準，主要以詞彙（vocabulary）和結構（structures）劃分。

Level 1 至 Level 3 出現的詞彙較淺顯，相對深的核心詞彙均配上中文解釋，節省讀者查找詞典的時間，以專心理解正文內容。在註釋的幫助下，讀者若能流暢地閱讀正文內容，就不用擔心這一本書程度過深。

Level 1 至 Level 3 出現的動詞時態形式和句子結構比較簡單。動詞時態形式以現在時（present simple）、現在時進行式（present continuous）、過去時（past simple）為主，句子結構大部分是簡單句（simple sentences）。此外，還包括比較級和最高級 （comparative and superlative forms）、可數和不可數名詞 （countable and uncountable nouns）以及冠詞（articles）等語法知識點。

Level 4 至 Level 6 出現的動詞時態形式，以現在完成時（present perfect）、現在完成時進行式（present perfect continuous）、過去完成時（past perfect continuous）為主，句子結構大部分是複合句（compound sentences）、條件從句（1st and 2nd conditional sentences）等。此外，還包括情態動詞（modal verbs）、被動形式（passive forms）、動名詞（gerunds）、

短語動詞（phrasal verbs）等語法知識點。

　　根據上述的語法範圍，讀者可按自己實際的英語水平，如詞彙量、語法知識、理解能力、閱讀能力等自主選擇，不再受制於學校年級劃分或學歷高低的約束，完全根據個人需要選擇合適的讀物。

② 怎樣提高閱讀效果？

　　閱讀的方法主要有兩種：一是泛讀，二是精讀。兩者各有功能，適當地結合使用，相輔相成，有事半功倍之效。

　　泛讀，指閱讀大量適合自己程度（可稍淺，但不能過深）、不同內容、風格、體裁的讀物，但求明白內容大意，不用花費太多時間鑽研細節，主要作用是多接觸英語，減輕對它的生疏感，鞏固以前所學過的英語，讓腦子在潛意識中吸收詞彙用法、語法結構等。

　　精讀，指小心認真地閱讀內容精彩、組織有條理、遣詞造句又正確的作品，着重點在於理解 "準確" 及 "深入"，欣賞其精彩獨到之處。精讀時，可充分利用書中精心設計的練習，學習掌握有用的英語詞彙和語法知識。精讀後，可再花十分鐘朗讀其中一小段有趣的文字，邊唸邊細心領會文字的結構和意思。

　　《Black Cat 優質英語階梯閱讀》中的作品均值得精讀，如時間有限，不妨嘗試每兩個星期泛讀一本，輔以每星期挑選書中一章精彩的文字精讀。要學好英語，持之以恆地泛讀和精讀英文是最有效的方法。

③ 本系列的練習與測試有何功能？

　　《Black Cat 優質英語階梯閱讀》特別注重練習的設計，為讀者考慮周到，切合實用需求，學習功能強。每章後均配有訓練聽、説、讀、寫四項技能的練習，分量、難度恰到好處。

聽力練習分兩類，一是重聽故事回答問題，二是聆聽主角對話、書信朗讀、或模擬記者訪問後寫出答案，旨在以生活化的練習形式逐步提高聽力。每本書均配有 CD 提供作品朗讀，朗讀者都是專業演員，英國作品由英國演員錄音，美國作品由美國演員錄音，務求增加聆聽的真實感和感染力。多聆聽英式和美式英語兩種發音，可讓讀者熟悉二者的差異，逐漸培養分辨英美發音的能力，提高聆聽理解的準確度。此外，模仿錄音朗讀故事或模仿主人翁在戲劇中的對白，都是訓練口語能力的好方法。

閱讀理解練習形式多樣化，有縱橫字謎、配對、填空、字句重組等等，注重訓練讀者的理解、推敲和聯想等多種閱讀技能。

寫作練習尤具新意，教讀者使用網式圖示（spidergrams）記錄重點，採用問答、書信、電報、記者採訪等多樣化形式，鼓勵讀者動手寫作。

書後更設有升級測試（Exit Test）及答案，供讀者檢查學習效果。充分利用書中的練習和測試，可全面提升聽、說、讀、寫四項技能。

4 本系列還能提供甚麼幫助？

《Black Cat 優質英語階梯閱讀》提倡豐富多元的現代閱讀，巧用書中提供的資訊，有助於提升英語理解力，擴闊視野。

每本書都設有專章介紹相關的歷史文化知識，經典名著更有作者生平、社會背景等資訊。書內富有表現力的彩色插圖、繪圖和照片，使閱讀充滿趣味，部分加上如何解讀古典名畫的指導，增長見識。有的書還提供一些與主題相關的網址，比如關於不同國家的節慶源流的網址，讓讀者多利用網上資源增進知識。

Contents

The story is recorded in full. 故事錄音

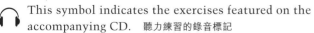 This symbol indicates the exercises featured on the accompanying CD. 聽力練習的錄音標記

BEFORE READING

Do you know these words?

coonskin cap
浣熊皮帽

wild cat
野貓

rifle
步槍

buckskin
trousers
鹿皮褲

buckskin jacket
鹿皮衣

racoon
浣熊

bear
熊

cattle
牛

8

2 **Which words do you think of when you read the name Davy Crockett? Circle them.**

building   forest   animals   books

pictures   rifle   mystery

Indians   ship

battle   fort   money

Mexico   soldier   castle   courageous

houses   Texas   England

America   church

king   strong   artist

3 **Do you know any other American stories? Which ones?**

4 **Do you like Western films?**

5 **What is your favourite Western film?**

6 **American Quiz**

What do you know about the United States of America?

a. How many states are there?

- [ ] 48
- [ ] 52
- [ ] 50

b. Name two famous Americans

- [ ] Abraham Lincoln
- [ ] Christopher Columbus
- [ ] Michael Jackson

c. Which is a typical American food?

- [ ] pasta
- [ ] hamburgers
- [ ] pizza

d. Name two big American cities

- [ ] Los Angeles, San Francisco
- [ ] New York, London
- [ ] Dallas, Montreal

e. Name a big American river

- [ ] Po
- [ ] Mississippi
- [ ] Thames

f. Name a favourite American sport

- [ ] rugby
- [ ] cricket
- [ ] basketball

CHAPTER ONE

Young Davy

 avy Crockett was born on August 17, 1787, in Greene County, Tennessee. In 1787 America was a new nation. Most of the territory was a wilderness [1].

Davy Crockett was born in the wilderness. His father, John Crockett, was an Irish immigrant [2]. He fought in the American Revolution. He was a very courageous [3] man.

Davy had a difficult childhood [4]. His family was poor. When he was twelve years old, his father said, "Davy, our family doesn't have any money. You must go to work as a cattle herder [5]."

"All right, father," said Davy. "I'm happy to help the family!"

Davy travelled four hundred miles on foot. He took cattle from Tennessee to other places. When he finished his work, he was far from home. He was lost [6]! Davy walked 400 miles. When he

1. **wilderness** : 荒野。
2. **immigrant** : 移民。
3. **courageous** : 勇敢的。
4. **childhood** : 童年。
5. **cattle herder** : 牛仔。
6. **lost** : 迷路的。

11

Davy Crockett

"An American Hero"

returned home, he was very tired.

Davy gave his father the money he made. His father was happy and said, "Thank you, Davy! Now we have some money for the winter months."

"I can do it again to help the family," Davy answered.

The years passed. Davy went to school when he had time. He spent most of his time hunting in the forest. He was the best shooter [1] and hunter [2] in Tennessee. He entered many shooting competitions [3] and won them all. He called his rifle "Old Betsy."

For years Davy hunted bears and other wild animals. He was a trapper [4]. Once he hunted one hundred bears in six months!

It was dangerous [5] to hunt bears but Davy was very courageous and strong. He knew the forest well. The Indians were his good friends. He ran fast and was a strong fighter. He fought with the wild cats of the mountains.

Some people said that one day Davy saw a racoon in a tree. He wanted to shoot it but the racoon saw him and said, "Wait a minute! Are you Davy Crockett?"

Davy answered, "Yes, I am!"

The racoon answered, "Then don't shoot! I'll come down from the tree." And the racoon came down from the tree!

---

1. **shooter**：射手。
2. **hunter**：獵人。
3. **competitions**：競賽。
4. **trapper**：設陷阱捕獸的人。
5. **dangerous**：危險的。

UNDERSTANDING THE TEXT

1 **Choose the correct answer.**

a. Davy Crockett was

☐ born in Greene County, Tennessee.
☐ an Irish immigrant.
☐ fought in the American Revolution.

b. When Davy was 12 years old

☐ he went to work as a hunter.
☐ he went to work as a cattle herder.
☐ he was an excellent shooter.

c. He went to work because

☐ he did not go to school.
☐ he wanted to travel far from his home.
☐ his family was poor.

d. He was the best shooter and hunter

☐ in America.
☐ in Tennessee.
☐ in Greene County.

e. Davy was a trapper. He hunted

☐ animals for their fur.
☐ animals for food.
☐ cattle.

f. It was dangerous to

☐ work as a cattle herder.
☐ hunt bears.
☐ live in Tennessee.

2 **Odd one out!**

Find the odd word out in each line.

a. shooter   hunter   mountains   trapper   fighter

b. happy   cattle   exhausted   courageous   strong

c. racoon   bear   wild cat   cattle   family

d. winter   rifle   summer   spring   autumn

Now use the odd words to complete these sentences.

1. Davy Crockett took .... from Tennessee to other places.

2. Wild cats live in the ....

3. Davy's .... was poor.

4. "Old Betsy" was Davy's ....

3 **In, on or at?**

You are a Journalist from the "Tennessee Gazette". You want to interview Davy. Fill in the gaps with the correct preposition of time（表示時間的介詞）**: in, on or at.**

Journalist: When were you born, Davy?

Davy: I was born .... August 17, 1787.

Journalist: When did you begin working as a cattle herder?

Davy: I began working .... April 1799.

Journalist: What do you do now?

Davy: I work as a hunter and trapper.

Journalist: Tell me about your day.

Davy: I get up ...... 5 a.m. every morning. .... Sundays I get up .... 7 a.m. .... midday I have lunch, and .... 6 p.m. I have dinner. I go to bed .... 9.30 p.m. I hunt wild animals .... the spring, summer and autumn. .... the winter, I sell the furs. .... Tuesdays I visit my friends.

CHAPTER TWO

The Creek Wars

*E*veryone *liked Davy Crockett.* He was always happy, with a big smile. He was *honest* and always *helped others.*

Davy was a tall man. He wore a coonskin cap, buckskin trousers and a buckskin jacket. He always carried his long rifle, "Old Betsy."

In 1806 Davy married Polly Finley. She was a school teacher.

They had two sons and a daughter: John, Joseph and Judith. After a few years, the Crockett family moved into the Tennessee hills. The Tennessee hills were near hostile [1] Indian country.

In 1812 the war between the United States and Britain began. The Mohawk and Creek Indians fought with the British against

---

1. **hostile**：含敵意的。

16

The Creek Wars

the Americans. *The American General Andrew Jackson organized a small army.* He wanted to fight the Creek Indians and the British.

Davy fought with this army. *He was a scout [1] because he knew the territory [2] well.* His work as a scout was very important. Davy travelled across Tennessee, the Mississippi Territory, Florida and Louisiana with General Jackson's army. *The Battle of New Orleans was a big victory for General Jackson and the Americans.* The war ended in 1814.

The British lost the war. The Indians lost their territories and went away. New American families settled [3] in the Tennessee hills.

At the end of the war Davy returned home to his family. Unfortunately, his wife Polly died. Life was again difficult for Davy. He worked as a trapper and took care of his three children.

After some time, Davy met Meg Mackinack. Meg's father was an American trapper. Her mother was a Cherokee Indian. Davy fell in love and married her. They had twins [4]. Davy named the twins George and Washington, in honour of [5] America's first president [6]. There were now five children in the Crockett family!

---

1. **scout** : 偵察者。
2. **territory** : 領土。
3. **settled** : 定居。
4. **twins** : 孿生兒。
5. **in honour of** : 紀念。
6. **president** : 總統。

UNDERSTANDING THE TEXT

1 **Are these sentences true (T) or false (F)? Correct the false ones.**

	T	F
a. Davy was honest and always helped others.	☐	☐
b. Davy's wife, Judith, was a school teacher.	☐	☐
c. Davy and Polly had three children.	☐	☐
d. The war between the United States and Britain began in 1806.	☐	☐
e. Davy fought with General Andrew Jackson's army.	☐	☐
f. The Americans lost the War of 1812.	☐	☐
g. After Polly died, Davy married Meg Mackinack.	☐	☐

2 **Find and circle the names of the four places where Davy travelled to during the War of 1812.**

M	F	P	X	C	I	O	H	Y	D	U	R	J	F
T	O	U	S	F	L	O	R	I	D	A	Q	L	G
B	E	G	A	P	K	G	X	T	U	K	B	O	T
F	C	N	L	R	O	E	A	M	K	Q	Y	U	D
J	T	K	N	P	H	Q	S	B	N	S	K	I	C
E	G	T	N	E	S	S	G	D	R	Y	B	S	G
A	V	M	I	S	S	I	S	S	I	P	P	I	H
X	F	R	E	A	P	S	Z	E	O	M	O	A	D
O	H	A	U	W	L	B	E	Y	E	D	B	N	K
Y	W	S	Q	O	C	T	O	E	C	J	W	A	Z

Now circle the four places on this map.

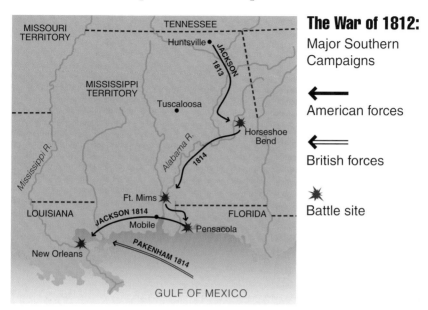

The War of 1812:

Major Southern Campaigns

← American forces

⇐ British forces

✳ Battle site

3 **Fill in the gaps with the correct preposition from the rifle.**

after   between   near   with   for   against   across

a. The hills were .... hostile Indian country.

b. In 1812 the war .... the United States and Britain began.

c. The Indians fought .... the British,
.... the Americans.

d. Davy travelled .... Tennessee, Florida and Louisiana.

e. Life was again difficult .... Davy.

f. .... a few years, the Crockett family went to the Tennessee hills.

a. ☐

b. ☐

c. ☐

d. ☐

e. ☐

f. ☐

Hunters and Trappers

Life in the American wilderness was exciting but it was not easy. There were wild animals, hostile Indians, no roads and no good maps.

Only strong, courageous men and women lived in the wilderness. These people were very independent. They were free spirits [1]. Davy Crockett and Jim Bowie (who you will read about later in the story) were hunters and trappers. They hunted wild animals for food and for their furs. Davy hunted bears and wild cats. Both were very dangerous animals.

Hunters and trappers had a difficult life. They lived alone in the high mountains for many months. Often it snowed and it was very cold. They did not have a home. They lived in a cave or in a small hut [2].

They hunted many animals: bears, wild cats, alligators, buffalo [3], beavers and others. Then they sold the furs at a trading post [4]. Trappers made good money by selling furs. Indians often worked as trappers too.

1. **free spirits** : 熱愛自由的人。
2. **hut** : 小木屋。
3. **buffalo** : 水牛。
4. **trading post** : 貿易站。

Animals that Trappers Hunted

Bobcat
紅貓

Beaver
海狸

Racoon
浣熊

Alligator
短吻鱷

Grizzly Bear
灰熊

BLACK CAT ENGLISH CLUB

The Commercial Press (Hong Kong) Ltd.
9/F, Eastern Central Plaza,
3 Yiu Hing Road, Shau Kei Wan,
Hong Kong.

BLACK CAT ENGLISH CLUB

Membership Application Form

BLACK CAT ENGLISH CLUB is for those who love English reading and seek for better English to share and learn with fun together.

Benefits offered: - Member Card

- Member badge, poster, bookmark

- Book discount coupon

- Black Cat English Reward Scheme

- English learning e-forum

- Surprise gift and more...

Simply fill out the application form below and fax it back to 2565 1113.

Join Now! It's FREE exclusively for readers who have purchased *Black Cat English Readers* !

The book(or book set) that you have purchased: _____

English Name:_____ (Surname) _____ (Given Name)

Chinese Name:_____

Address: _____

Tel: _____ Fax: _____

Email:_____

Sex: ❑ Male   ❑ Female   (Login password for e-forum will be sent to this email address.)

Education Background: ❑ Primary 1-3   ❑ Primary 4-6   ❑ Junior Secondary Education (F1-3)

❑ Senior Secondary Education (F4-5)   ❑ Matriculation

❑ College   ❑ University or above

Age: ❑ 6 - 9   ❑ 10 - 12   ❑ 13 - 15   ❑ 16 - 18   ❑ 19 - 24   ❑ 25 - 34

❑ 35 - 44   ❑ 45 - 54   ❑ 55 or above

Occupation: ❑ Student   ❑ Teacher   ❑ White Collar   ❑ Blue Collar

❑ Professional   ❑ Manager   ❑ Business Owner   ❑ Housewife

❑ Others (please specify: _____)

As a member, what would you like **BLACK CAT ENGLISH CLUB** to offer:

❑ Member gathering/ party   ❑ English class with native teacher   ❑ English competition

❑ Newsletter   ❑ Online sharing   ❑ Book fair

❑ Book discount   ❑ Others (please specify: _____)

Other suggestions to **BLACK CAT ENGLISH CLUB**:

_____

Please sign here: _____

(Date: _____)

1 Complete the description of a trapper who lives in the American wilderness. Use the words in the hut.

> free spirit
> difficult   wilderness
> courageous   money   hut
> independent   mountains
> trading post
> cave   animals   furs

An American trapper lived in the .... His life was .... He was .... and .... He was a .... In the .... he lived in a .... or a small .... .The trapper hunted many .... He sold the .... at a .... He made good .... by selling furs.

23

CHAPTER THREE

Davy Enters Politics

Davy and his big family wanted to live in Lawrenceburg, Tennessee. Davy bought a covered wagon [1]. He and his family put all of their things inside the wagon: chairs, tables, beds, clothing and many other things. Four strong horses pulled the big wagon. Davy and his wife sat in the front. The five children were inside the wagon.

After travelling in the forests and mountains, they arrived in Lawrenceburg. This was their new home. Davy opened a mill [2]. He was a hard worker. His neighbours liked him. Everybody in Lawrenceburg liked him.

He became a representative [3] of the town government. He was

1. **covered wagon** :
2. **mill** : 磨坊。
3. **representative** : 代表。

Davy Crockett

"An American Hero"

very popular. People liked listening to his stories about hunting bears and about the War of 1812. They admired his honesty and courage.

Davy was different from other politicians [1]. His speeches were never boring [2]. His message was clear. He spoke to the people in simple language. He dressed like them too. He always wore his coonskin cap and his buckskin trousers and jacket.

One day at an important meeting he didn't know what to say. So he looked at the people and said, "Today, I am like a man trying to drink water from an empty barrel [3]! I'll tell you a funny story and then we can go home!"

Davy soon became a representative of the government of Tennessee. Now he was in politics. He helped his people in many ways.

At first, many politicians laughed at Davy because he never wore a suit [4]. After some time, these politicians admired and respected him. Davy was an honest man. Everyone believed what he said.

His buckskin jacket had two big pockets. In his right pocket Davy had a bottle of whisky. When he met his friends, he gave them some whisky! In those days, it was common to give some whisky to friends.

1. **politicians** : 政治家。
2. **boring** : 無趣的。
3. **barrel** :
4. **suit** : 西服。

UNDERSTANDING THE TEXT

1 **Circle the correct answer.**

   a. Davy and his family went to *Lawrenceville / Lawrenceburg*, Tennessee.

   b. He and his *friends / family* put all of their things inside a covered *wagon / boat*.

   c. In Lawrenceburg, Davy became a *representative / number* of the town government.

   d. People liked listening to his *stories / songs* about hunting bears and the War of 1812.

   e. Davy's speeches were never *interesting / boring* and his message was *clear / confusing*.

   f. Many *politicians / hunters* laughed at Davy because he never wore a suit.

   g. When he met his friends, he gave them some *food / whisky.*

2 **Fill in the gaps with the correct "wh" words:** where, what, which, who, when **and complete the crossword.**

   1. .... wanted to live in Lawrenceburg?

   2. .... did the family put inside the covered wagon?

   3. .... did Davy and his wife sit?

   4. In .... pocket was the bottle of whisky?

   5. .... did Davy give his friends some whisky?

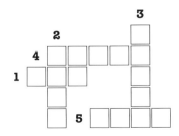

3 Find the hidden word

Take the letters you need to make the words that match the pictures. Put the remaining letter in the box below.

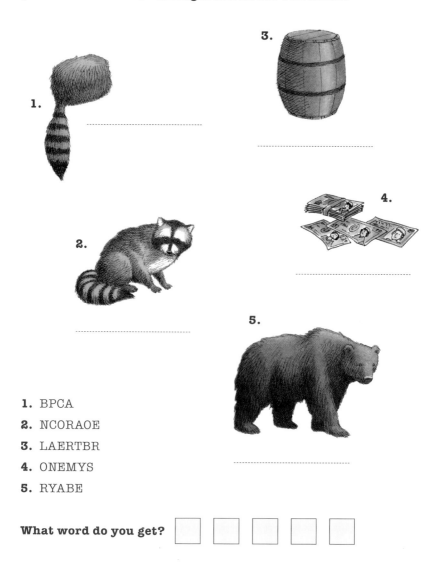

1. BPCA
2. NCORAOE
3. LAERTBR
4. ONEMYS
5. RYABE

What word do you get? ☐ ☐ ☐ ☐ ☐

4 Match these words with their meaning.

1. twins ☐
2. shooter ☐
3. trapper ☐
4. wilderness ☐
5. hunter ☐
6. Tennessee ☐

a. person who kills animals for food
b. an American state
c. person who catches wild animals to sell their furs
d. two children born of the same mother, at the same time
e. person who shoots with a rifle or gun
f. land that is not cultivated, where only wild animals live

CHAPTER FOUR

Congressman[1] Crockett

Davy often went to Nashville, the capital of Tennessee. In Nashville, he worked for the Tennessee Government. One day, while Davy was working for the Tennessee Government, a big flood[2] destroyed his mill. This was terrible! He was very unhappy because he lost a lot of money.

When he returned to Lawrenceburg, he started a new type of work. This time he made barrels! He sold these barrels in New Orleans, Louisiana. New Orleans was about 400 miles away.

To go to New Orleans, he travelled on the Mississippi River. He had a big boat to carry the barrels. It was difficult to travel on the Mississippi River. There were many dangers.

1. **Congressman**：國會議員。
2. **flood**：洪水。

Davy Crockett

"An American Hero"

One day his boat had a bad accident [1]. It began to sink [2]. Davy almost drowned [3]. He lost his boat and his barrels but he didn't lose his life!

In 1827 there was a big election [4] in Tennessee. Davy Crockett became a United States Congressman! This was a great honour for him. In the United States Congress, he represented the people of Tennessee. He travelled to Washington D. C., the capital of the United States.

Davy was very happy to be a Congressman. He wanted to help his people. There were Congressmen who wanted to take land away from the Fox Indians of Tennessee. Davy protected these Indians. He fought against dishonest [5] Congressmen.

"The Fox Indians cannot live without their land!" Davy shouted. "I must defend all the American people of Tennessee: the whites and the Indians!"

After months of hard work, Davy was not able to help the Fox Indians. The U. S. Congress made a law that took away land from the Indians. Davy was very angry. He hated injustice [6]. In 1835 he left the U. S. Congress!

1. **accident** : 意外。
2. **sink** : 下沉。
3. **drowned** : 淹死。
4. **election** : 選舉。
5. **dishonest** : 不誠實的。
6. **injustice** : 非正義。

UNDERSTANDING THE TEXT

 1 Fill in the gaps with the correct words from the covered wagon.

accident

barrels   Indians   boat

New Orleans   flood

Nashville   law   defended

Mississippi River

Congressman

a. The capital of Tennessee is ....

b. A .... destroyed Davy's mill.

c. Davy made .... and sold them in ....

d. He travelled to New Orleans on the ....

e. One day, he had an .... and lost his ....

f. In 1827 Davy became a United States ....

g. In Congress he .... the people of Tennessee.

h. The United States Congress made a .... that took away land from the ....

2 Match the characters with their descriptions.

1. Polly Finley
2. John Crockett
3. Andrew Jackson
4. John, Joseph and Judith
5. Davy Crockett
6. George and Washington
7. Meg Mackinack

a. Davy's second wife
b. U.S. Congressman
c. Davy Crockett's children
d. Davy's father
e. American General
f. school teacher
g. Irish immigrant
h. Davy's first wife
i. trapper
j. Davy Crockett's twins

3 Must and mustn't

Look at this example from Chapter 4:

"I *must* defend all the American people of Tennessee: the whites and the Indians!"

We use must **and** mustn't **to say that it is necessary to do or not to do something.**

Use must/mustn't **to fill in the gaps in the following sentences.**

a. Davy said "I .... start a new type of work."

b. He .... be careful on the Mississippi River because it's dangerous.

c. We .... build a big boat to carry the barrels.

d. "I .... drown," said Davy when his boat began to sink!

e. Now that Davy is a Congressman he .... go to Washington D.C.

f. "The U.S. Congress .... take away the land from the Indians," shouted Davy.

g. Dishonest people .... be Congressmen.

4 **Davy wrote a letter to his father, John Crockett. What did he tell him? Put the verbs in the Past Simple tense** (過去時).

Dear Father,

Last year a flood (destroy) .... my mill. Meg and I (be) .... very unhappy. We (lose) .... a lot of money.

Last month I (have) .... a bad accident on the Mississippi River. I (lose) .... my boat and I almost (drown) ....

On November 10th, I (become) .... a United States Congressman. I (go) .... to Washington D.C. I (work) .... there. I (fight) .... to protect the Indians, but I (be) .... not able to help them.

The U.S. Congress (make) .... a law. This law (take) .... away land from the Fox Indians.

I (be) .... very angry. I (leave) .... Congress!

Love, Davy

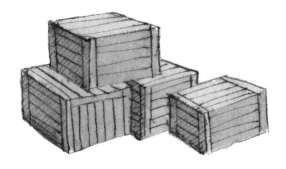

CHAPTER FIVE

Texas!

In 1835 the Crockett family left Tennessee. They put all their things in a big covered wagon and they travelled for many days. They finally arrived in Texas.

At that time, Texas belonged to [1] Mexico. At first, the Mexican Government was happy with the American settlers. They cultivated [2] the land.

Many settlers went to Texas because land was very cheap. With a little money, a settler bought a lot of land. More and more Americans went to Texas. There were about 20,000 American settlers in eastern Texas. There were only 5,000 Mexicans! The Mexican Government did not like this. There were too many Americans!

Mexico decided to close its borders [3]. The American settlers were very angry.

In 1834, the Mexican General Santa Anna became the dictator [4]

---

1. **belonged to** : 屬於。
2. **cultivated** : 耕種。
3. **borders** : 邊界。
4. **dictator** : 獨裁者。

Davy Crockett
"An American Hero"

of Mexico. He was a cruel man. He sent his soldiers to the Mexican border. He did not want American settlers to enter Mexico.

By 1834, there were more than 30,000 Americans living in Texas. They wanted Texas to be an independent American state! They did not want to live under a cruel Mexican dictator.

Davy and his family now lived in Texas. They were happy in their new home. They wanted the independence of Texas too.

One day Davy heard that the Mexicans wanted to attack Fort Alamo. The Alamo was a Spanish church and fort near San Antonio, in western Texas. There were Texans and American soldiers at Fort Alamo. There were also women and children at the fort. War was in the air [1]! Davy knew he must fight for the independence of Texas.

1. **in the air**：即將發生。

UNDERSTANDING THE TEXT

 1 **Choose the correct answer.**

a. In 1835 the Crockett family
- [ ] went to Washington D.C.
- [ ] left Tennessee and went to Texas.
- [ ] left Tennessee and went to San Antonio.

b. At that time Texas
- [ ] was an American state.
- [ ] was an independent territory.
- [ ] belonged to Mexico.

c. American settlers went to Texas
- [ ] because land was very cheap.
- [ ] because they liked General Santa Anna.
- [ ] because there was a war in the United States.

d. General Santa Anna was
- [ ] the Mexican king.
- [ ] the Mexican dictator.
- [ ] the American General in Texas.

e. In 1834 there were
- [ ] 30,000 Americans living in Texas.
- [ ] 5,000 Americans living in Texas.
- [ ] no Americans living in Texas.

f. Fort Alamo was
- [ ] a Mexican city.
- [ ] a Spanish church and fort.
- [ ] the capital of Texas.

Have fun with this crossword puzzle!

ACROSS

1. Davy was born here

2.

3. Fort

4.

5. General Jackson's first name

DOWN

6. American settlers bought land here

7. A city in Louisiana: New .....

8.

9. Past Simple of verb "to shoot"

10. to go to a place to live a long time

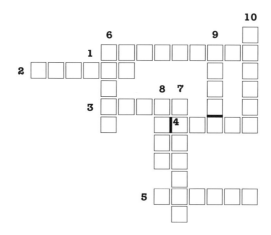

3 Possessive adjectives（物主形容詞）

Look at this sentence:

They put all *their* things in a big covered wagon.
Their is a possessive adjective.

The other possessive adjectives are: my, your, his, her, its, our.
Fill in the gaps with the correct possessive adjectives.

a. Davy named .... rifle "Old Betsy".

b. Meg was happy in .... new home.

c. "Mexico is .... country!" said General Santa Anna.

d. "This is .... new land," said Davy to the Indians.

e. Mexico decided to close .... borders.

f. "......... covered wagon is full!" said the Crockett family.

4 Listen to the first three paragraphs of Chapter 5 and fill in the missing words. If necessary listen twice.

In 1835, the Crockett family .... Tennessee. They
.... all their things in a big covered wagon and they
travelled for many .... They finally arrived in Texas.
At that time, ... belonged to Mexico. At first, the
Mexican Government was .... with the American
settlers. They cultivated the land.

Many settlers .... to Texas because land was very cheap.
With a little money, a settler bought a lot of .... More and
more Americans .... to Texas. There were about 20,000
American settlers in eastern Texas. There .... only 5,000
Mexicans! The Mexican Government did not .... this. There
were too .... Americans!

The American Frontier[1]

In the 18th century America was a very young nation. Only small parts of the East were settled by white people.

On the American continent *there were forests, lakes, rivers, valleys, tall mountains and deserts.* But *there were very few people.* American Indians lived on this enormous[2] continent.

Every year thousands of white settlers explored[3] the American frontier. These *settlers moved from the East to the West* in their covered wagons. Some settled near the rivers, others in the valleys. Most of them were farmers. Some were cowboys with their cattle. *Towns and cities grew quickly.*

1. **Frontier** : 邊境。
2. **enormous** : 巨大的。
3. **explored** : 考察。

In our story, American settlers moved to Texas. Texas was a Mexican territory. The settlers fought for the independence of Texas, which soon became an American state!

This is how America grew. *Year after year, new states became part of America.* Today there are 50 states in the United States.

1 It's the summer of 1835. You are a journalist in Washington D.C. You work for the "Washington Pioneer". Your friend in New York must write a short article about the American frontier but he doesn't have any information! Send him a telegram and give him some information—remember, make it short!

Look at the highlighted sections of The American Frontier and prepare the telegram.

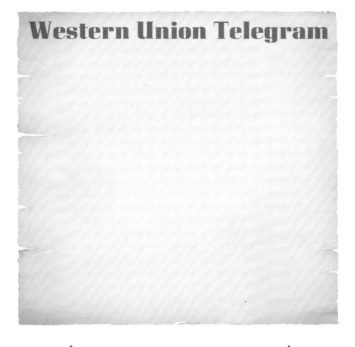

Western Union Telegram

CHAPTER SIX

Fort Alamo

avy asked other American settlers to go to the Alamo with him. Few men wanted to fight. But this did not stop Davy! He and 15 men decided to go to the Alamo. They were ready to fight the Mexicans.

There were 112 men at the Alamo. Colonel William Travis of the U.S. Army was the commander. William Travis was a young colonel [1]. He was only 27 years old. He was a lawyer [2]. He entered the U.S. Army to fight for the independence of Texas. One day, Colonel Jim Bowie and 30 men arrived at the fort. Jim Bowie was a tall, strong man. He was a hunter and trapper.

"Good evening, Colonel Travis," said Colonel Bowie. "I have a message for you from General Sam Houston. Here is the letter."

Colonel Travis opened it. He read it aloud:

1. **colonel**：上校。
2. **lawyer**：律師。

Fort  Alamo

You must destroy the Alamo and come with my army! General Santa Anna will attack the Alamo soon.

Gen. Sam Houston

"What!" said Colonel Travis. "I don't want to destroy the Alamo. I want to defend it."

Colonel Bowie said, "We cannot defend the Alamo. We must have more men."

In February 1836, Davy Crockett and his men arrived at the fort. Colonel Travis was happy to see them. He asked Davy and his men to defend the Alamo.

"We don't have many men," said Colonel Travis. "We must ask for more soldiers. I am sending a messenger [1] to General Fannin. He can send us more soldiers."

Davy said, "My men and I want to defend the Alamo. We are hunters and trappers. Our long rifles can shoot at a great distance [2]."

Davy Crockett and Jim Bowie became good friends. Together they repaired the walls of the fort. They cleaned the rifles and the cannons [3]. They were ready for the battle.

1. **messenger** : 信使。
2. **at a great distance** : 遠距離。
3. **cannons** :

UNDERSTANDING THE TEXT

1 **Are these sentences true (T) or false (F)? Correct the false ones.**

	T	F
a. Davy asked other American settlers to go to the Alamo with him.	☐	☐
b. All the men wanted to fight.	☐	☐
c. Colonel William Travis was the commander of the Alamo.	☐	☐
d. Colonel Jim Bowie had a message from General Sam Houston.	☐	☐
e. The message said, "Defend the Alamo!"	☐	☐
f. Davy and his fifteen men arrived at the Alamo in February 1836.	☐	☐
g. Davy and Colonel Travis became good friends.	☐	☐

2 **Who said or wrote it?**

Decide which of the four characters made the following statements:

1. "I have a message for you from General Houston." ☐
2. "Our long rifles can shoot long distances." ☐
3. "You must destroy the Alamo and come with my army." ☐
4. "I want to defend it!" ☐
5. "We cannot defend the Alamo. We must have more men." ☐
6. "We must ask for more soldiers." ☐
7. "We are hunters and trappers." ☐
8. "I am sending a messenger to General Fannin." ☐

a. Colonel Jim Bowie

b. Colonel William Travis

c. Davy Crockett

d. General Sam Houston

3 **Negative sentences**（否定句）.

Look at these negative sentences:

This *did not* stop Davy!
I don't want to destroy the Alamo.

Make the following sentences negative.

a. Davy asked other men to go to Texas.

....

b. They arrived at the fort in December.

....

c. They were ready to fight.

....

d. He can send us more soldiers.

....

Now make these negative sentences affirmative（把否定句變為肯定句）.

e. He did not receive the message.

....

f. We cannot shoot well.

....

g. They did not clean the rifles.

....

h. We don't want to defend the fort.

....

4 When Davy arrived at the Alamo he wrote a letter to his wife. What did he tell her? Use the words in the envelope to complete the sentences.

return   battle
soldiers
defend   fifteen   commander
a lot of   friend   young
children   walls

Dear Meg,

I didn't .... home because I decided to .... the Alamo. I found .... men who want to help me. There are .... courageous .... here.

Colonel Travis is the .... He is a .... man.

Colonel Jim Bowie is a good .... Yesterday we cleaned the rifles and the cannons. We repaired the .... of the fort.

We are ready for the ....

Love to you and the ....,

Davy

5 Find the names of the two Colonels and two Generals mentioned in Chapter 6. Circle them in red.

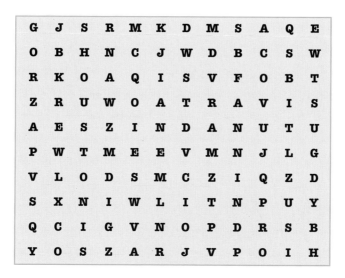

G	J	S	R	M	K	D	M	S	A	Q	E
O	B	H	N	C	J	W	D	B	C	S	W
R	K	O	A	Q	I	S	V	F	O	B	T
Z	R	U	W	O	A	T	R	A	V	I	S
A	E	S	Z	I	N	D	A	N	U	T	U
P	W	T	M	E	E	V	M	N	J	L	G
V	L	O	D	S	M	C	Z	I	Q	Z	D
S	X	N	I	W	L	I	T	N	P	U	Y
Q	C	I	G	V	N	O	P	D	R	S	B
Y	O	S	Z	A	R	J	V	P	O	I	H

a. Who wrote the letter?

....

b. Who does the messenger go to, asking for more soldiers?

....

c. Who is the commander?

....

d. Who arrived at the fort with 30 men?

....

Jim Bowie
and the Bowie Knife

J im Bowie was a great American adventurer and hero. He was born in 1796 in Kentucky. When he was 6 years old, his family went to live in Louisiana.

When Jim was 18 years old, he hunted bears and alligators! He also captured wild horses. He was a tall, strong man. He loved

adventures and danger. Jim Bowie was never afraid of anything.

In 1831 Jim Bowie settled in Texas. He bought some land near San Antonio. He married the Vice-Governor's daughter, Maria. He and his family loved Texas. At the Alamo, Bowie fought for the independence of his new home, Texas.

He and his brother invented the famous Bowie knife. It was built in a special way to protect the hand. It was an excellent knife for hunting and fighting. This

knife became very popular in the United States and Britain. Every pioneer [1] and cowboy had a Bowie knife. It is still used today.

---

1. **pioneer**：拓荒者。

1 Are these sentences true (T) or false (F)? Correct the false ones.

		T	F
a.	Jim Bowie was an American General.	☐	☐
b.	He was born in Kentucky in 1769.	☐	☐
c.	He captured wild horses.	☐	☐
d.	Bowie was a very courageous man.	☐	☐
e.	Jim and his brother invented the Bowie knife.	☐	☐
f.	He and his family bought land in Tennessee.	☐	☐
g.	Bowie did not fight at the Alamo.	☐	☐

2 How many words can you find in the word river? ....
Circle them in red.

knifefoadventurefkherotwoalligatorsrstrongtallaopdangersifmlpioneerbucowboyuwithhorses

Have fun with this crossword puzzle!

ACROSS

1. He was a ...., strong man.

2. Jim's wife.

3.

4.

5. Jim was born in 1796 in ....

DOWN

6. Jim was a .... and an adventurer.

7.

8. Jim's family went to live in ....

9.

CHAPTER SEVEN

The Red Flag[1]

ne morning a messenger arrived. "I have a message
from General Fannin: He is very sorry. He cannot
send any soldiers. You must leave the Alamo now.
General Santa Anna is near. You are all in danger."

"We don't want to leave the Alamo," said Jim Bowie. "We
want to fight for the independence of Texas. How many soldiers
has General Santa Anna got?"

"He's got about 4,000 soldiers!" said the messenger.

Colonel Bowie looked at Colonel Travis. "We must speak to our
men," said Colonel Travis. "Yes," said Bowie, "we must speak to
them."

Colonel Travis called his 187 men. He said sadly, "General
Santa Anna is coming to attack us. He has about 4,000 soldiers
and lots of ammunition[2]. We have only 187 men and little

---

1. flag :

2. **ammunition**：彈藥，軍火。

Davy Crockett

"An American Hero"

ammunition. We have little food and water. Remember, there are women and children in the fort."

Then he marked a line on the ground with his sword [1]. "Those who want to fight for the independence of Texas, cross this line! The others can leave the fort and go home."

There was a very long silence. Davy thought about his wife and his five children. Then he thought about the independence of Texas and the American settlers. He thought about a new American state: Texas!

1. **sword** :

 # The Red Flag

All the men crossed the line. These men wanted to defend the Alamo. These men wanted an independent Texas.

On February 23, 1836, General Santa Anna and his army arrived. He sent a messenger to the Alamo. He wanted the people in the Alamo to leave. The men in the Alamo answered with a cannon shot! They did not want to leave. They did not want to return to the United States. General Santa Anna was furious. He showed a red flag. The red flag meant: No Prisoners! He wanted to kill everyone in the fort!

On February 24, 1836, General Santa Anna's army attacked the Alamo. The Mexicans had an enormous cannon. It shot a cannon

Davy Crockett
"An American Hero"

ball that damaged a wall of the fort.

Davy said, "We must destroy that cannon. It can destroy the walls of our fort."

That night, Davy and Jim Bowie left the fort. They went to the Mexican camp. Everyone in the camp was sleeping. Davy and Jim silently passed behind the two Mexican guards. They put mud [1] and stones [2] in the enormous cannon. Then they returned to the Alamo.

The next morning, the Mexican army used the cannon. It exploded [3]! The Mexicans were very surprised. Their enormous cannon was destroyed.

The men in the Alamo killed many Mexican soldiers. The Mexican army attacked many times during the day. But the Texans and the American army defended the Alamo.

1. **mud** : 泥。

2. **stones** :

3. **exploded** : 爆炸。

56

UNDERSTANDING THE TEXT

1 **Fill in the gaps with the correct words from the fort.**

mud and stones   sword
cannon   messenger
crossed   exploded
leave   independence
attack   danger   4,000
ground   ammunition

a. One morning a .... arrived.

b. You must .... the Alamo now. You are all in
....

c. We want to fight for the .... of Texas.

d. General Santa Anna is coming to .... us.

e. He has about .... soldiers and lots of
....

f. He marked a line on the .... with his
....

g. All the men courageously .... the line.

h. The Mexicans had an enormous ....

i. Davy and Jim put .... in the enormous cannon.

j. The next morning, the cannon ....

2 Write the names of the characters under the pictures. Then choose the words from the coonskin cap to describe these characters. Some words can be used twice.

honest   friendly

trapper   commander

young   courageous

dictator   tall   strong

lawyer   Mexican   cruel

hunter

a.

....
....
....
....

b.

....
....
....
....

c.

....
....
....
....

d.

....
....
....
....

Now write a sentence describing each character.

a. Davy was ....

b. Jim ....

c. General Santa Anna ....

d. Colonel Travis ....

58

Object Pronouns (賓語人稱代詞)

There are two types of pronouns: Subject Pronouns (主語人稱代詞)
(I you he she it we they**) and Object Pronouns (**me you him her it
us them**).**

Look at this example from Chapter 7:

"We must speak to our men," said Colonel Travis.
"Yes," said Bowie, "We must speak to *them*."
Them refers to the men.

We use object pronouns as the direct or indirect object of a verb
(動詞的直接或間接賓語)**. Substitute the words in italics with the**
correct object pronoun.

a. There were American soldiers at Fort Alamo.
A lot of Mexican soldiers wanted to destroy *Fort Alamo.*
....

b. There are few soldiers at General Fannin's camp.
General Fannin cannot send *the soldiers* to the fort.
....

c. Davy did not know Jim Bowie.
Davy met *Jim* at the Alamo.
....

d. General Santa Anna showed the red flag. Colonel Travis said:
"The red flag is a message for *me and the other people*," said
Colonel Travis.
....

e. There were some good rifles at the Alamo.
Jim gave a rifle to Davy and said, "This rifle is for *Davy.*"
....

f. Davy's wife is hungry.
The food and water are for *Davy's wife.*
....

CHAPTER EIGHT

The Massacre [1]

The days passed. The battle continued. There was little food and ammunition. The men in the Alamo fought courageously. Nothing stopped them!

After twelve days of fighting, General Santa Anna sent all of his army to attack the fort. On March 6, at 5 a.m., the Mexican bugles [2] *played the "Deguello."* The "Deguello" was a war song. It meant "Death for everyone." The women and children at the fort were tired and afraid. The situation was desperate [3].

The men at the Alamo heard the "Deguello". They understood the message. *General Santa Anna's army attacked the fort from all sides.* It was a terrible battle. Everyone was shooting. Cannon balls were flying. A lot of men were injured [4]. Some men were

---

1. **massacre** : 屠殺。

2. **bugles** :

3. **desperate** : 極嚴重的。
4. **injured** : 受傷的。

killed. Davy and the other men defended the fort. They sent back the Mexicans twice.

The third time, *the Mexicans entered the fort*. They killed many people: men, women and children. *It was a massacre!*

Jim Bowie was a great fighter. He was a strong man and he was never afraid. *He fought with his famous Bowie knife.* He killed many Mexican soldiers. At the end of the day, *three Mexican soldiers killed Jim Bowie.*

Davy and the other men fought until the end. They killed many enemy soldiers. It was a desperate battle. Four Mexican soldiers killed Davy with a long knife. The tall, courageous trapper fell to the ground!

Only two women and two children were alive [1] after the massacre. But General Santa Anna did not kill them. These women and children returned home.

The Mexicans burnt the bodies of the dead people. It was a big victory for General Santa Anna. He lost 1,544 men at the Alamo.

On April 21, 1836, General Sam Houston and his army attacked General Santa Anna. *General Houston made General Santa Anna prisoner. Santa Anna then signed [2] a treaty [3]. This treaty said that Texas was independent!* Everyone remembered

---

1. **alive** : 活着的。
2. **signed** : 簽署
3. **treaty** : 條約。

Davy Crockett

"An American Hero"

Davy Crockett, Jim Bowie, Colonel Travis and the other men. They died for the independence of Texas. *Texas became a state of the United States in 1845.*

All through his life, Davy Crockett did what he believed was right. With his honesty and determination [1], he became a national hero. Davy Crockett was the perfect example of the American free spirit. In his life story Davy wrote, "I leave this rule for others when I'm dead:

Always be sure you're right—then go ahead [2]!"

1. **determination** : 決心。   2. **go ahead!** : 開始做。

UNDERSTANDING THE TEXT

1 Choose the correct answer.

a. The men at the Alamo fought

☐ for 12 days.
☐ from March to April.
☐ for 2 days.

b. The Mexican "Deguello" was

☐ a long rifle.
☐ an enormous cannon.
☐ a war song.

c. On March 6, 1836, the Mexicans entered the Alamo

☐ but they did not fight.
☐ and there was a massacre.
☐ and Texas became independent.

d. Davy Crockett, Jim Bowie, Colonel Travis and the other men

☐ left the Alamo and returned home.
☐ fought until the end.
☐ went to fight with General Sam Houston's army.

e. On April 21, 1836, General Houston and his army

☐ destroyed Santa Anna's army and made him prisoner.
☐ defended the Alamo.
☐ declared the independence of Texas.

f. General Santa Anna signed a treaty

☐ with Colonel Travis.
☐ that said that Texas was independent.
☐ with General Fannin.

2 **Here you have the Infinitives of the verbs** (動詞原形) **that appear in Chapter 8. Go back and underline the Past Simple** (過去時) **of the verbs below. Then write them next to their infinitives.**

a.	fight	....	**h.**	return	....
b.	send	....	**i.**	lose	....
c.	mean	....	**j.**	make	....
d.	hear	....	**k.**	sign	....
e.	understand	....	**l.**	remember	....
f.	fall	....	**m.**	die	....
g.	do	....	**n.**	become	....

3 **Listen to the first two paragraphs of Chapter 8 and fill in the missing words.**

The .... passed. The battle continued. There was little

.... and ammunition. The .... in the Alamo

fought courageously. Nothing stopped ....!

After twelve days of fighting, General Santa Anna ....

all of his army to attack the fort. On March 6, at 5 a.m., the

Mexican bugles played the "Deguello." The "Deguello" was a

war .... It meant "Death for everyone." The women

and children at the .... were tired and afraid. The

situation was desperate.

Davy Crockett State Park

Near the city of Lawrenceburg in Tennessee, there is a big State Park called Davy Crockett State Park. It was created in 1959 in honour of America's great hero.

There is an interesting museum about Davy Crockett, his life and his times.

In this State Park there is a beautiful lake, a big park, an enormous sports area and there are camping grounds [1].

1. **camping grounds** : 露營地。

1 Let's complete Davy!

C _ O _ _ K _ _ _ A _

B _ _ _ S _ _ _ _ J _ _ _ E _

_ N _ _ _ _

T _ _ U _ _ _ _ _

R _ _ _ _ _

B _ _ _ _ _

EXIT TEST

CONTEXT

 Match the correct parts of the sentences:

1. In the 18th century	**a.** moved to the West.
2. Only small parts of the East	**b.** farmers and cowboys.
3. American Indians	**c.** America was a very young nation.
4. Thousands of settlers	**d.** fifty states in the U.S.A.
5. Most of them were	**e.** were settled by white people.
6. Today there are	**f.** lived on this enormous continent.

COMPREHENSION

 Circle the correct word.

Davy Crockett was born in ¹*Tennessee/Kentucky*. He was a hunter and ²*farmer/trapper*. He was a ³*tall/short* man and wore a coonskin ⁴*cap/jacket*. He always carried his long ⁵*pistol/rifle* called 'Old Betsy.' He fought with General Andrew ⁶*Travis/Jackson* in the War of 1812 against the ⁷*Spanish/British*. Everyone liked Davy because he was friendly and ⁸*rich/honest*. He became a U.S. ⁹*Congressman/Colonel* and travelled to Washington D.C. The U.S. Congress made a law that took ¹⁰*money/land* away from the Indians. Davy was very ¹¹*happy/angry* and he left the U.S. Congress. In 1835 ¹²*Davy/the Crockett family* went to Texas. At that time Texas belonged to ¹³*Britain/Mexico*. Davy and other men wanted to fight for the the independence of ¹⁴*America/Texas*. Davy, Jim Bowie, Colonel Travis and others decided to ¹⁵*destroy/defend* the Alamo. They fought courageously ¹⁶*against/with* the Mexican Army After 12 days of ¹⁷*fighting/waiting* General Santa Anna's army attacked the Alamo. Davy, Jim Bowie, Colonel Travis and the other man were ¹⁸*injured/killed*. In 1845 ¹⁹*Texas/Tennessee* became an American state. Davy's courage, honesty and determination made him a national hero.

67

GRAMMAR

3 **Make the following sentences negative.**

a. Davy hunted bears in the winter.

....

b. The politicians laughed at Davy.

....

c. You are ready to go to Louisiana.

....

d. The general sent more soldiers to the fort.

....

e. He was a good scout.

....

Now make these sentences affirmative.

f. They did not attack the fort. ....

g. Jim did not clean the rifles. ....

h. You cannot shoot well. ....

i. She was not ready to go. ....

j. We didn't see the bear in the forest. ....

4 **Substitute the words in italics with the correct object pronoun: *me, you, him, her, it, us, them.***

1. Davy cannot see *the soldiers.*

....

2. The letter is for *me and Jim.*

....

3. The food is for Davy's wife.

....

4. They want to defend *the Alamo.*

....

5. Colonel Travis did not know *General Santa Anna.*

....

6. Davy gave some flowers to Meg and said, "These flowers are for Meg."

....

5 **What is your favourite part of the story?** ....

6 **Write a few sentences about Davy Crockett and Jim Bowie.**

"An American Hero"
Davy Crockett

KEY TO
THE EXERCISES
AND EXIT TEST

KEY TO THE EXERCISES

BEFORE READING

Page 9 Exercise 2-5
Open answers

Page 10 Exercise 6
a. 50
b. Abraham Lincoln, Michael Jackson
c. hamburgers
d. Los Angeles, San Francisco
e. Mississippi
f. basketball

CHAPTER 1
Page 14 Exercise 1
a. born in Greene County, Tennessee
b. he went to work as a cattle herder
c. his family was poor
d. in Tennessee
e. animals for their fur
f. hunt bears

Page 15 Exercise 2
a. mountains
b. cattle
c. family
d. rifle

1. cattle
2. mountains
3. family
4. rifle

Page 15 Exercise 3
on, in, at, On, at, At, at, at, in, In, On

CHAPTER 2
Page 18 Exercise 1
a. T
b. F – Davy's wife, Polly was a school teacher.
c. T
d. F – The war between the United States and Britain began in 1812.
e. T
f. F – The Americans won the War of 1812.
g. T

Page 18 Exercise 2

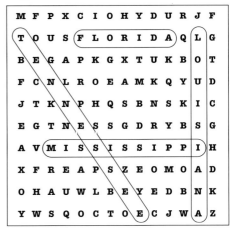

Page 28 Exercise 3
1. cap
2. racoon
3. barrel
4. money
5. bear

What word do you get? BETSY

Page 29 Exercise 4
1d - 2e - 3c - 4f - 5a - 6b

CHAPTER 4

Page 33 Exercise 1
a. Nashville
b. flood
c. barrels, New Orleans
d. Mississippi River
e. accident, boat
f. Congressman
g. defended
h. law, Indians

Page 34 Exercise 2
1f, h - 2d, g - 3e - 4c - 5b, i - 6 j - 7a.

Page 34 Exercise 3
a. must
b. must
c. must
d. mustn't
e. must
f. mustn't
g. mustn't

Page 35 Exercise 4
destroyed, were, lost, had, lost,
drowned, became, went, worked,
fought, was, made, took, was, left

CHAPTER 5

Page 39 Exercise 1
a. left Tennessee and went to Texas
b. belonged to Mexico
c. because land was very cheap
d. the Mexican dictator
e. 30,000 Americans living in Texas
f. a Spanish church and fort

Page 19 Exercise 3
a. near
b. between
c. with, against
d. across
e. for
f. After

Page 20 Exercise 4
a6 - b3 - c4 - d5 - e2 - f1.

HUNTERS AND TRAPPERS

Page 23 Exercise 1
wilderness, difficult, courageous,
independent, free spirit, mountains,
cave, hut, animals, furs, trading
post, money

CHAPTER 3

Page 27 Exercise 1
a. Lawrenceburg
b. family, wagon
c. representative
d. stories
e. boring, clear
f. politicians
g. whisky

Page 27 Exercise 2
1. Who
2. What
3. Where

Page 40 Exercise 2

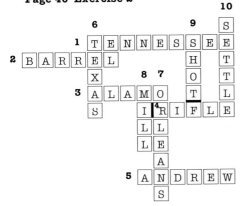

Page 41 Exercise 3
a. his; **b.** her; **c.** our; **d.** your; **e.** its;
f. Our

Page 41 Exercise 4
left, put, days, Texas, happy, went,
land, went, were, like, many

CHAPTER 6

Page 46 Exercise 1
a. T
b. F – Few men wanted to fight.
c. T
d. T
e. F – The message said, "Destroy
the Alamo!"
f. T
g. F – Davy and Jim Bowie became
good friends.

Page 46 Exercise 2
1a - 2c - 3d - 4b - 5a - 6b - 7c - 8b.

Page 47 Exercise 3
a. Davy didn't ask other men to go
to Texas.
b. They didn't arrive at the fort in
December.
c. They weren't ready to fight.
d. He can't send us more soldiers.
e. He received the message.
f. We can shoot well.
g. They cleaned the rifles.
h. We want to defend the fort

Page 48 Exercise 4
return, defend, fifteen, a lot of,
soldiers, commander, young, friend,
walls, battle, children

Page 49 Exercise 5

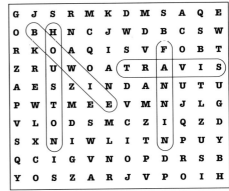

a. General Sam Houston
b. Colonel Travis
c. Colonel Travis
d. Colonel Jim Bowie

JIM BOWIE AND THE BOWIE KNIFE

Page 51 Exercise 1
a. F – Jim Bowie was an American
adventurer and hero
b. F – He was born in Kentucky in
1796
c. T
d. T
e. T
f. F – He and his family bought land
near San Antonio
g. F – Bowie fought at the Alamo.

Page 51 Exercise 2
knife, adventure, hero, alligators,
strong, tall, dangers, if, pioneer,
cowboy, with, horses

Page 52 Exercise 3

CHAPTER 7

Page 57 Exercise 1
a. messenger
b. leave, danger
c. independence
d. attack
e. 4,000, ammunition
f. ground, sword
g. crossed
h. cannon
i. mud and stones
j. exploded

Page 58 Exercise 2
a. Davy Crockett: honest, friendly, trapper, young, courageous, tall, strong
b. Jim Bowie: trapper, hunter, young, courageous, tall, strong
c. General Santa Anna: dictator, cruel, Mexican
d. Colonel Travis: commander, lawyer, young, courageous

Page 59 Exercise 3
a. it
b. them
c. him
d. us
e. you
f. her

CHAPTER 8

Page 63 Exercise 1
a. for 12 days
b. a war song
c. and there was a massacre
d. fought until the end
e. destroyed Santa Anna's army and made him prisoner
f. that said that Texas was independent

Page 64 Exercise 2
a. fought
b. sent
c. meant
d. heard
e. understood
f. fell
g. did
h. returned
i. lost
j. made
k. signed
l. remembered
m. died
n. became

Page 64 Exercise 3
days, food, men, them, sent, song, fort

DAVY CROCKETT STATE PARK

Page 66 Exercise 1

1. 1. c **2.** e **3.** f **4.** a **5.** b **6.** d

2. 1. Tennessee **2.** trapper **3.** tall **4.** cap **5.** rifle **6.** Jackson
7. British **8.** honest **9.** U.S. Congressman **10.** land **11.** angry
12. the Crockett family **13.** Mexico **14.** Texas **15.** defend
16. against **17.** fighting **18.** killed **19.** Texas

3. a. Davy didn't hunt bears in the winter.
b. The politicians didn't laugh at Davy.
c. You aren't ready to go to Louisiana.
d. The general didn't send more soldiers to the fort.
e. He wasn't a good scout.
f. They attacked the fort.
g. Jim cleaned the rifles.
h. You can shoot well.
i. She was ready to go.
j. We saw the bear in the forest.

4. 1. them **2.** us **3.** her **4.** it **5.** him **6.** you

5. Open answer.

6. Open answer.

notes

notes

notes

notes

Name of Book: Davy Crockett
Told by: Gina D. B. Clemen
Editors: Rebecca Raynes, Monika Marszewska
Design: Nadia Maestri
Illustrations: Alfredo Belli
Edition: ©1998 Black Cat Publishing
an imprint of Cideb Editrice, Genoa, Canterbury

系 列 名：Black Cat 優質英語階梯閱讀 · Level 1
書　　名：大衛克羅傳
顧　　問：Angeli Lau
責任編輯：傅　伊
封面設計：張　毅
出　　版：商務印書館（香港）有限公司
　　　　　香港筲箕灣耀興道3號東滙廣場8樓
　　　　　http://www.commercialpress.com.hk
印　　刷：中華商務彩色印刷有限公司
　　　　　香港新界大埔汀麗路36號中華商務印刷大廈
版　　次：2003年10月第1版第3次印刷
　　　　　© 2003 商務印書館（香港）有限公司
　　　　　ISBN 962 07 1636 1
　　　　　Printed in Hong Kong

Black Cat English Readers

BLACK CAT ENGLISH CLUB

Membership Application Form

BLACK CAT ENGLISH CLUB is for those who love English reading and seek for better English to share and learn with fun together.

Benefits offered:
- *Membership Card*
- *Member badge, poster, bookmark*
- *Book discount coupon*
- *Black Cat English Reward Scheme*
- *English learning e-forum*
- *Surprise gift and more...*

Simply fill out the application form below and fax it back to 2565 1113.

Join Now! It's FREE exclusively for readers who have purchased *Black Cat English Readers* !

The book(or book set) that you have purchased: _____

English Name: _____ (Surname) _____ (Given Name)

Chinese Name: _____

Address: _____

Tel: _____ Fax: _____

Email: _____
(Login password for e-forum will be sent to this email address.)

Sex: ❏ Male   ❏ Female

Education Background: ❏ Primary 1-3   ❏ Primary 4-6   ❏ Junior Secondary Education (F1-3)
❏ Senior Secondary Education (F4-5)   ❏ Matriculation
❏ College   ❏ University or above

Age: ❏ 6 - 9   ❏ 10 - 12   ❏ 13 - 15   ❏ 16 - 18   ❏ 19 - 24   ❏ 25 - 34
❏ 35 - 44   ❏ 45 - 54   ❏ 55 or above

Occupation: ❏ Student   ❏ Teacher   ❏ White Collar   ❏ Blue Collar
❏ Professional   ❏ Manager   ❏ Business Owner   ❏ Housewife
❏ Others (please specify: _____)

As a member, what would you like **BLACK CAT ENGLISH CLUB** to offer:
❏ Member gathering/ party   ❏ English class with native teacher   ❏ English competition
❏ Newsletter   ❏ Online sharing   ❏ Book fair
❏ Book discount   ❏ Others (please specify: _____)

Other suggestions to **BLACK CAT ENGLISH CLUB**:

_____

Please sign here: _____

(Date: _____)